I0817172

Forest Babies

Mary Elizabeth Salzmann

Consulting Editor, Diane Craig, M.A./Reading Specialist

Sandcastle

An Imprint of Abdo Publishing
abdobooks.com

abdobooks.com

Published by Abdo Publishing, a division of ABDO, PO Box 398166, Minneapolis, Minnesota 55439.

Printed in the United States of America, North Mankato, Minnesota

052019
092019

Design: Christa Schneider, Mighty Media, Inc.
Production: Mighty Media, Inc.
Cover Photograph: Shutterstock Images
Interior Photographs: Shutterstock Images (all)

Library of Congress Control Number: 2018966940

Publisher's Cataloging-in-Publication Data
Names: Salzmann, Mary Elizabeth, author.
Title: Forest babies / by Mary Elizabeth Salzmann
Description: Minneapolis, Minnesota : Abdo Publishing, 2020 | Series: Animal babies
Identifiers: ISBN 9781532119583 (lib. bdg.) | ISBN 9781532174346 (ebook)
Subjects: LCSH: Forest animals--Juvenile literature. | Animal babies--Juvenile literature. | Forest animals--Behavior--Juvenile literature.
Classification: DDC 599.139--dc23

SandCastle™ Level: Emerging

SandCastle™ books are created by a team of professional educators, reading specialists, and content developers around five essential components—phonemic awareness, phonics, vocabulary, text comprehension, and fluency—to assist young readers as they develop reading skills and strategies and increase their general knowledge. All books are written, reviewed, and leveled for guided reading and early reading intervention programs for use in shared, guided, and independent reading and writing activities to support a balanced approach to literacy instruction. The SandCastle™ series has four levels that correspond to early literacy development. The levels are provided to help teachers and parents select appropriate books for young readers.

EMERGING • BEGINNING • TRANSITIONAL • FLUENT

Contents

Forest Babies

Can you find these baby forest animals in this book?

baby bear

baby chipmunk

baby deer

baby fox

baby moose

baby rabbit

baby raccoon

baby wolf

The baby rabbit has fur.

The baby moose has fur.

The baby deer has fur.

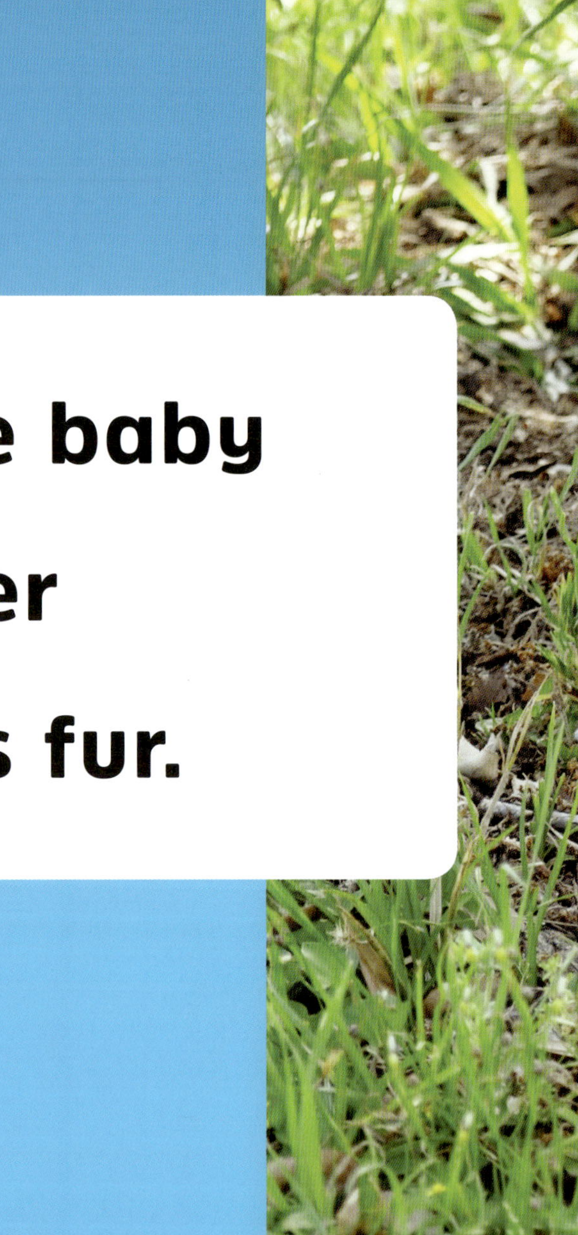

The baby wolf has fur.

The baby raccoon has fur.

The baby chipmunk has fur.

The baby
bear
has fur.

The baby fox has fur.

What Else Did You See?

branch

grass

leaves

rock

Index

Teacher's Guide

ATOS: 0.8 GRL: A Word Count: 40

High-Frequency Words

has The

Content Words

baby, bear, chipmunk, deer, fox, fur, moose, rabbit, raccoon, wolf

Before Reading

- Tell students that the title of the book is *Forest Babies*.
- Summarize the content of the book.
- Have students look through the book. Ask them what they see in the pictures.
- Choose a few new vocabulary words. Have students predict what letter each word starts with. Then have them find the words in the book.

After Reading

Ask students questions about the book's content, such as:

- What animals did you see in the book?
- How could you tell that the animals were in the forest?
- Have you ever been to a forest? What was it like?
- What other animals would you like to read about?